THE PAPER SKYSCRAPER

The technology of materials

David Drew

Illustrated by
Terry Denton and Peter Gouldthorpe

GINN

Brass Movie Ticket

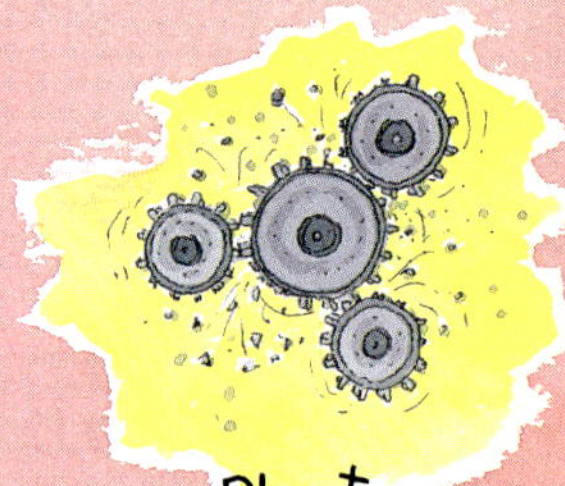
Plaster Gears

China Axe

Contents

Styrofoam Scuba Gear

Plastic Car Engine

Concrete Money

Silk Garbage Can

Leather Screwdriver

Clay Ocean Liner

Plastic Lighthouse

Glass Bicycle

Stone Dog Collar

Paper Airplane

Steel Shoes

Rubber Nails

Plastic Iron

China Bus

Plaster Shovel

Silk Ruler

Glass Tennis Racket

How to use this book

1. This book is about the materials that things are made of, such as iron, or plastic, or clay.

Use the clues on each page to guess the mystery material.

2. Then think what it would be like if things were made differently.

What would happen if skyscrapers were made of paper? Or if books were made of concrete?

Cardboard Boat

Rubber Sails

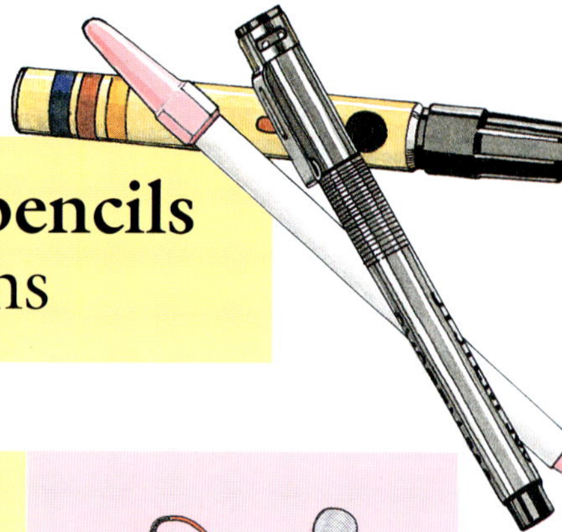

If you find it in pencils
but not in pens

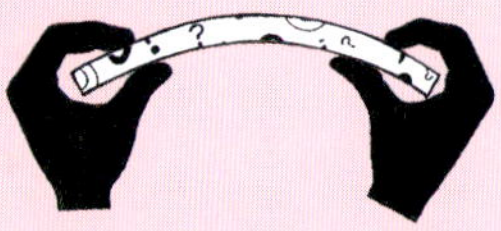

and you can bend it

it absorbs water

it does not con-duct electricity

it burns but it does not melt

it rots if it is wet

you can carve it with a chisel

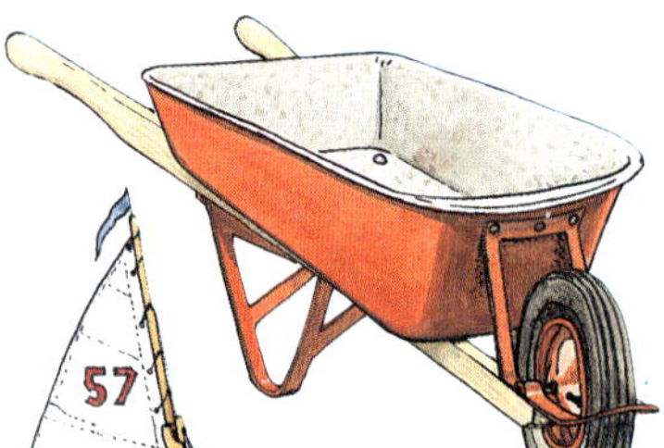

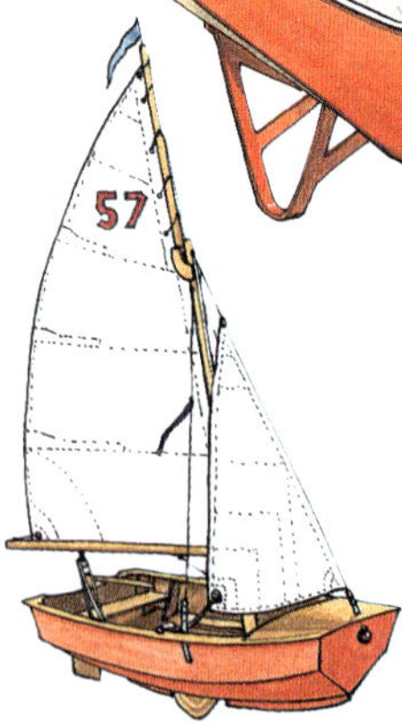

and you can join it with nails

then it must be ...

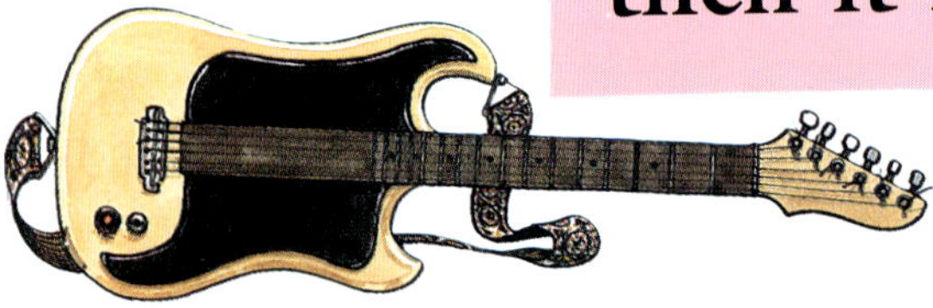

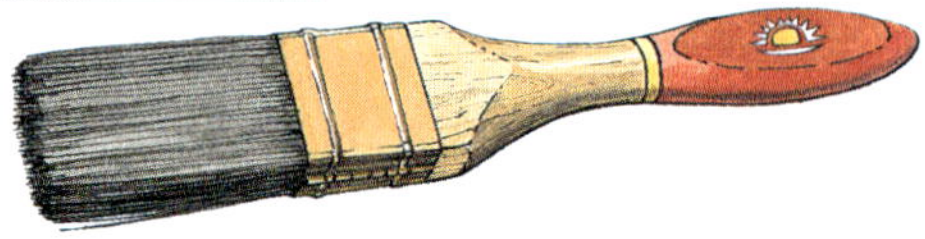

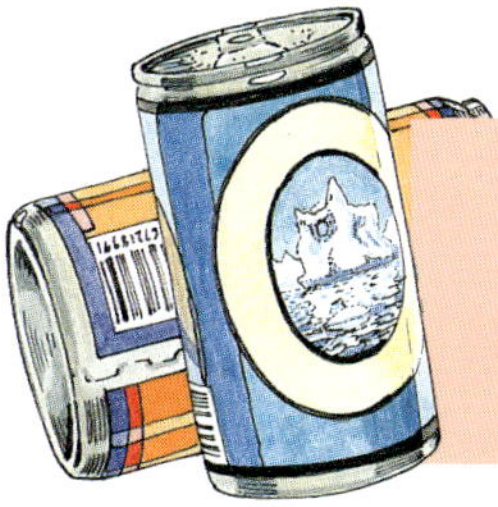

If you find it in cans but not in bottles

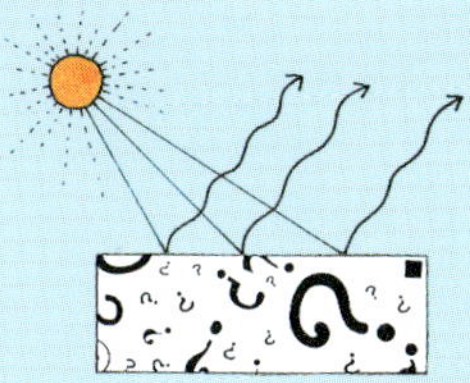

and it reflects heat

you can bend and fold it

it conducts electricity

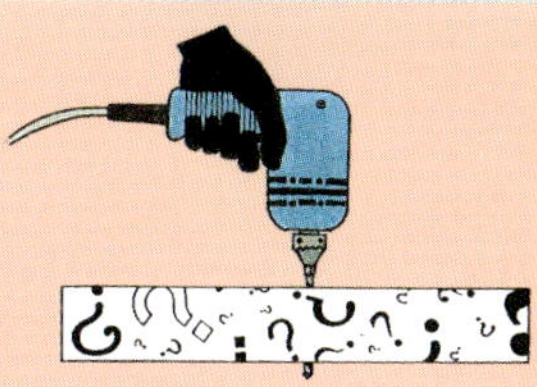

you can drill holes in it

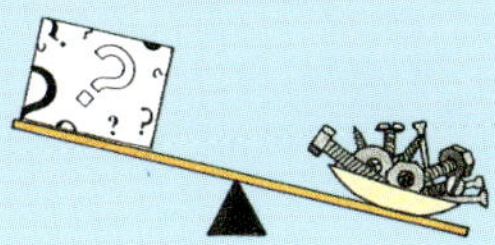

it is lighter than steel

you can join it with nails

and you can recycle it

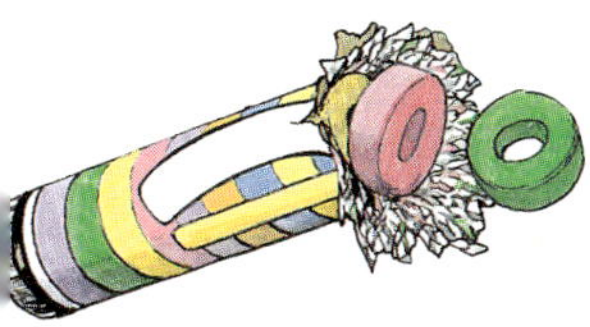

then it must be ...

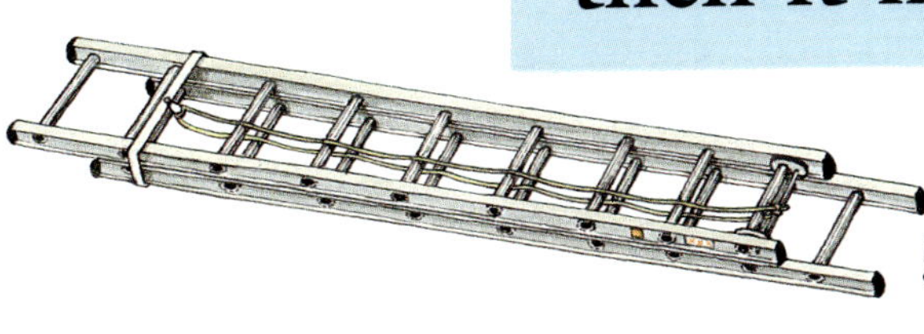

If you find it in shoes but not in laces

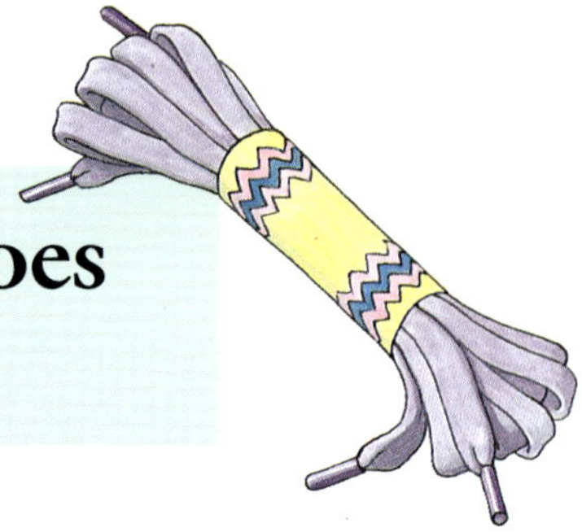

and it sinks in water

it is unbreakable

you can cut it with a knife

it is flammable

you can join it with glue

it is waterproof

and you can stretch it

then it must be ...

If you find it in needles
but not in thread

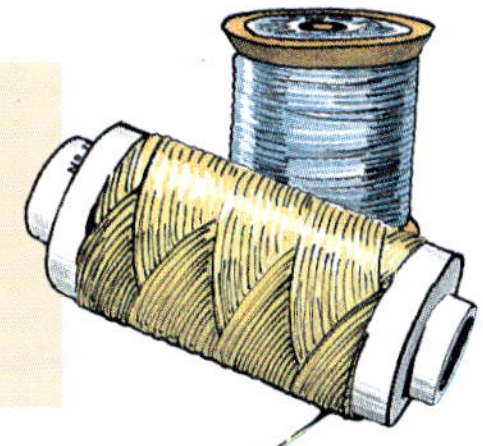

and it is magnetic

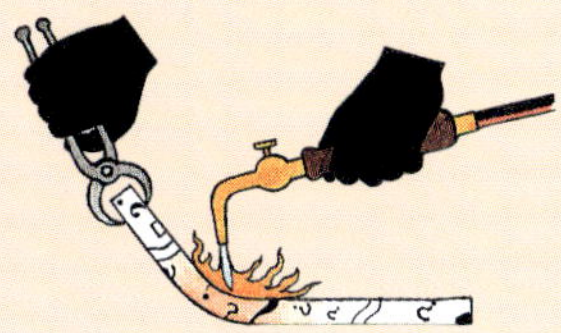

you can bend it if it is red hot

it sinks in water

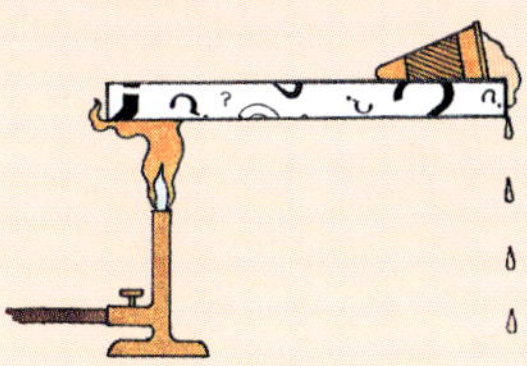

it conducts heat

it chimes when it is struck

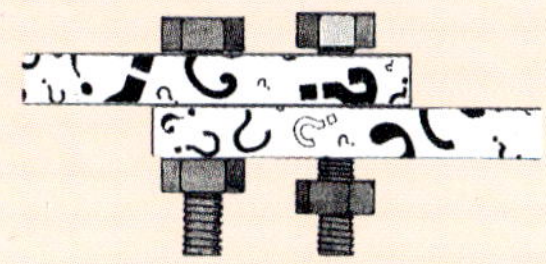

you can join it with bolts

and it rusts when it is wet

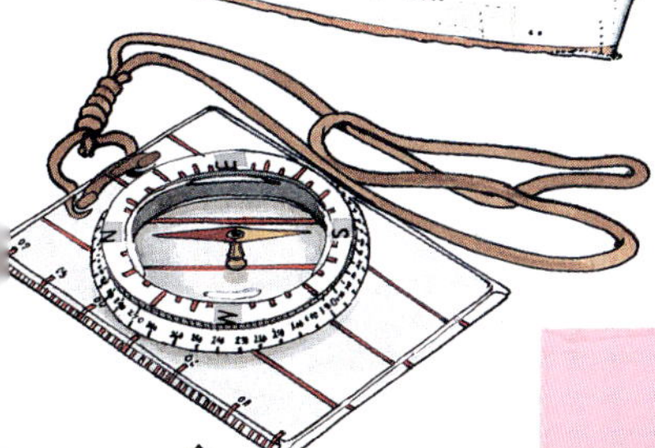

then it must be ...

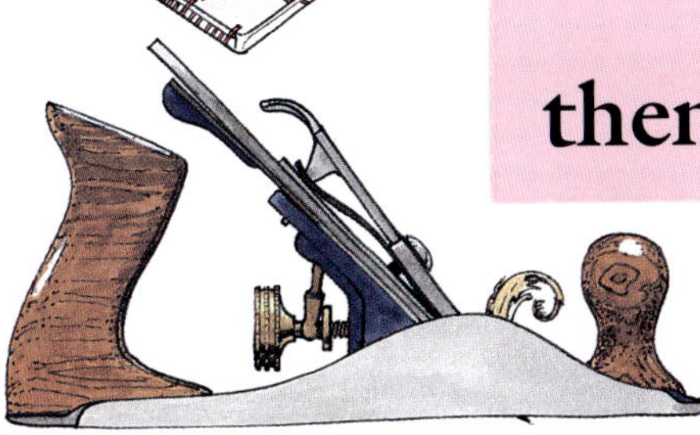

But what if ...

we had rubber bridges?

aluminum shoes?

steel fishbowls?

or steel raincoats?

rubber kettles?

wooden saws?

And what if ...

we had aluminum hoses?

wooden
blankets?

steel
toothbrushes?

rubber cars?

aluminum
sunglasses?

or wooden barbecues?

If you find it in cameras
but not in film

and it sinks in water

you can bend a thin thread of it

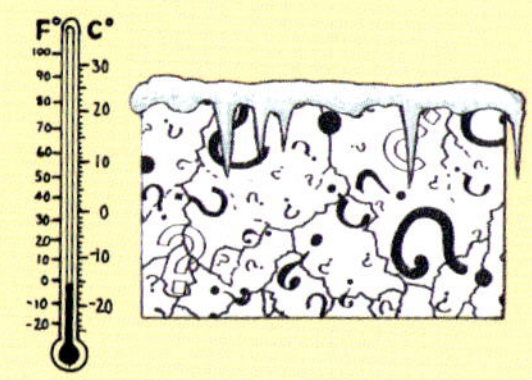

it will crack if you freeze it

it is brittle and fragile

it is waterproof

you can shape it if it is red hot

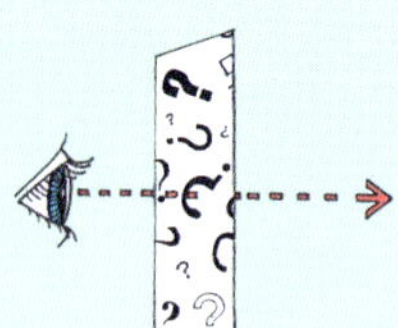

and it is transparent

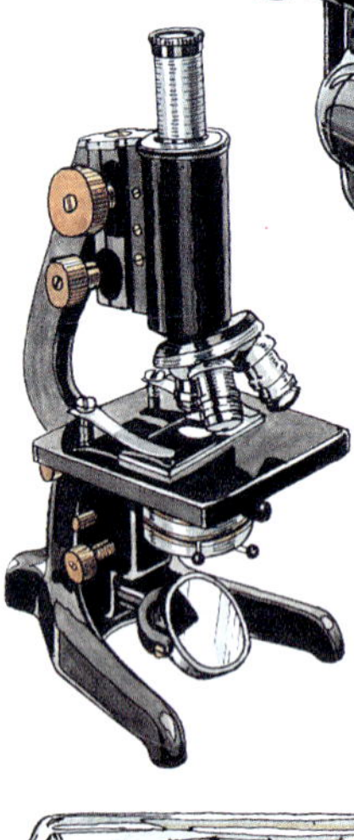

then it must be ...

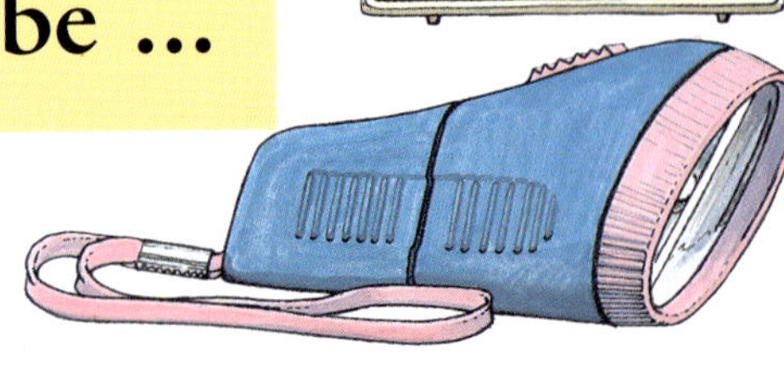

If you find it in faxes but not in phones

and you can't stretch it

you can shape it when it is wet

you can fold it

it absorbs water

it is translucent

you can join it with glue

and you can tear it

then it must be ...

If you find it in buildings
but not in bricks

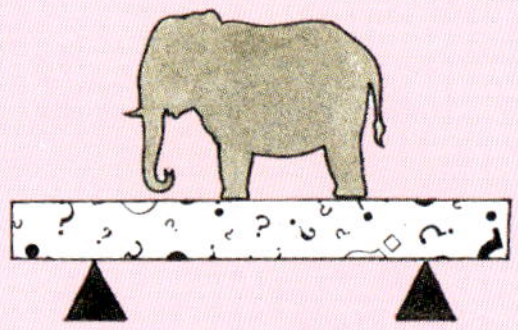

and it can bear heavy weights

it is brittle and fragile

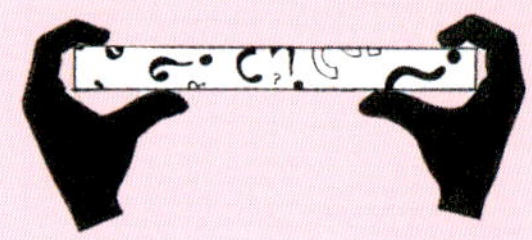

you can't bend it

it conducts heat slowly

it absorbs water

it can be shaped before it sets

and you can join it with cement

then it must be ...

If you find it in computers
but not in typewriters

and it can be hammered flat

you can scratch it with a needle

you can cut it with a knife

it conducts electricity

you can bend and twist it

you can join it by hammering

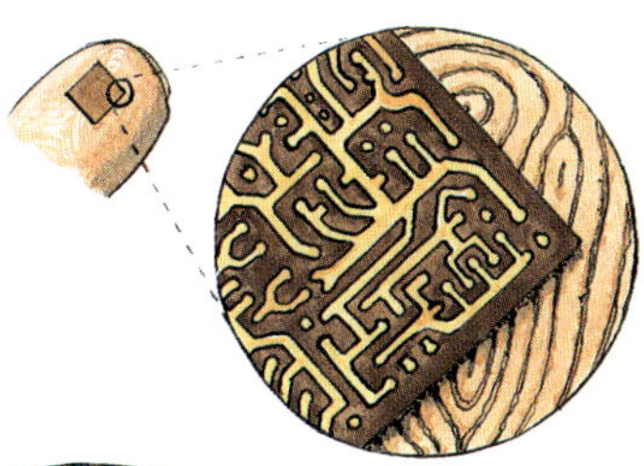

and it does not rust or tarnish

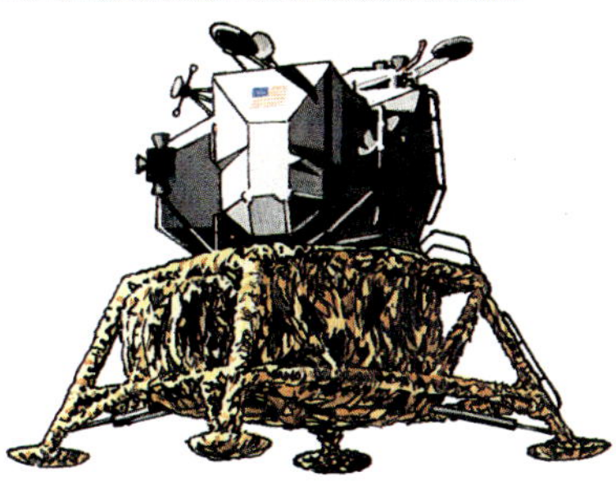

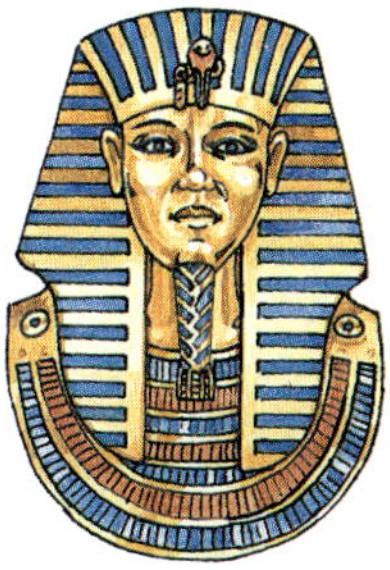

then it must be ...

But what if ...

we had concrete kites?

glass bridges?

paper clocks?

gold newspapers?

or glass towels?

concrete knives and forks?

And what if …

we had gold highways?

concrete saxophones?

glass fillings?

or paper bathtubs?

paper keys?

gold hammers?

Glossary

absorb *(v.)* to take into itself [*a sponge absorbs water*].
bear *(v.)* to support, carry.
brittle *(adj.)* easily cracked or shattered.
conduct *(v.)* to move something from one part to another [*copper wire conducts electricity*].
facsimile machine *(n.)* a machine for sending words or pictures by means of telephone lines.
fax *(n.)* abbrev. of *facsimile machine* (q.v.)
flammable *(adj.)* capable of being set on fire.
fragile *(adj.)* easily broken.
material *(n.)* the elements of which something is made [*plastic is a softer material than iron*].
tarnish *(v.)* to lose its shine, to become dull [*brass is quickly tarnished*].
translucent *(adj.)* capable of allowing some light to pass through.
transparent *(adj.)* capable of being looked through.

Index